BIRD-WATCHING WHERE DORSET MEETS DEVON

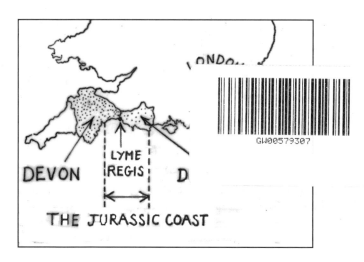

A PRACTICAL INSIGHT INTO ORNITHOLOGY

Written & Illustrated by

DR COLIN DAWES BSc PhD

ISBN 0-9520112-2-0

Printed by Charles Wheadon. Herald Graphics, Reading.

PREFACE

The variety of birds to be seen in a given area depends largely on the range of habitats the area has to offer. This publication takes in cliffs, rocky sea-shores, open countryside, dense woodland, heathland, rivers and an estuary. Each of these habitats is frequented by particular species of birds and is within easy reach of Lyme Regis, a small and largely unspoilt coastal resort situated on the Dorset/Devon coastal border.

Many of the birds described here may be unfamiliar to a reader who spends most of his life in a city or its suburbs. And it is for such a reader that this booklet has, in the main, been written. Typical birds that visit every English garden, such as the **Robin** *Erithacus rubecula* and **Blackbird** *Turdus merula* are largely ignored with the immediate exception of the **Song Thrush** *Turdus philomelos,* illustrated below. This bird was dear to the heart of probably England's greatest ornithologist: John Gould. He was born in Lyme Regis during the year 1804 and this booklet is dedicated to his memory following the bicentenary of his birth.

The booklet also doubles up as a sort of celebration following the establishment of superb bird-watching facilities on the Axe Estuary six miles to the west of Lyme Regis. The work was carried out by East Devon District Council under the guidance of its Local Nature Reserves Officer Fraser Rush and with the support of the Axe Vale and District Conservation Society. Its chairman is Donald Campbell, author of "The Encyclopedia of British Birds"* and voluntary warden of The Undercliffs National Nature Reserve west of Lyme Regis. His encouragement and help in getting this publication off the ground is gratefully acknowledged.

Colin Dawes 2005

The Song thrush. *Turdus philomelos.* It is said that twelve year old John Gould was lifted up by his father to see a nest of this species in Lyme Regis and that from that moment he decided to devote his life to the study of birds.

* Published by Parragon, ISBN 0 75256-564-8

CONTENTS

BACKGROUND

Lyme Regis & the History of Ornithology 4
The Jurassic Connection 7

PREPARING FOR A BIRD-WATCHING TRIP

Sensible precautions. 8
What to take 8
What to expect 8
Code of Conduct. 9
What to wear 9
Where to stay & where to go. 9
Map of the Dorset/Devon Coastal border 10
Map of Lyme Regis. 13

BIRDS OF LYME REGIS

The Herring Gull 11
Other Gulls & similar birds 16
Cormorants. 18
Purple Sandpipers & other wading birds 19
Rock Pipits & Wagtails 22
Dippers. 24
Buzzards 25
Stonechats 26
Wheatears 27

FURTHER AFIELD

Rare Coastal Birds: Ravens & Peregrine Falcons 28
Trinity Hill: Nightjars & other heathland birds 30
Stonebarrow Hill & Dartford Warblers 33
Lamberts Castle & the Marshwood Vale 34
Exploring the Undercliff. 35
Map of the Axe Estuary. 40
Exploring the Axe Estuary. 41
Birds of Beer Head 53

BACK TO LYME REGIS

Taking a boat trip 55

BACKGROUND

LYME REGIS AND THE HISTORY OF ORNITHOLOGY

In tracing the modern history of ornithology Lyme Regis is certainly a worthy starting point if only because it was the birthplace of **John Gould**. He is familiar to every Australian ornithologist but many English people have never heard of him in spite of their well known love of birds. In Australia Gould is known as "The Father of Ornithology" and it is to him that we owe the introduction of the Budgerigar (an Australian species) to this country.

Gould travelled extensively in his relentless pursuit of new species of birds that could be depicted by his team of artists, for whom he made annotated sketches. His artists included his wife and Edward Lear, better known for his nonsense poetry, and Joseph Wolf, arguably the greatest 19th century bird artist if we ignore **John James Audubon** (1785 - 1851) of America. The works produced by these artists stimulated a world-wide interest in birds, helping towards their conservation.

To the American bird-lover Audubon is a hero. America has its Audubon Society and Australia its John Gould Society but England has no corresponding human figurehead. Its laudable organisation devoted to the conservation of birds is simply known as the Royal Society for the Protection of Birds: The R.S.P.B.

Gould and Audubon lived at a time when more birds were around than today. These men witnessed what might be called (by the European) "The Golden Years of Ornithology" – the first fifty years of the 19th century, before the Passenger Pigeon was wiped out with the Bison in America and before the Great Bustard was exterminated in England.

The ornithological legacy of John Gould is a monumental series of bird prints. Two of these prints are on permanent display in the Philpot Museum of Lyme Regis; not bad considering that Gould's work cost a fortune in his own time and were produced under subscription on behalf of a patronage which included Queen Victoria, several foreign princes and a president of the United States of America.

Hardly anything is known about Gould's association with Lyme Regis. He left the town before he was a teenager. His father was the head gardener of a property situated on the high ground to the west of the town. Gould senior later became the head gardener of Windsor Castle. He took his son with him. The subsequent history of John Gould has been written up in a delightful publication by Maureen Lambourne entitled "John Gould - Bird Man" (Osberton Publications, 1987). The book is illustrated with many fine examples of Gould's artwork.

The gardens as John Gould knew them as a boy have changed beyond recognition in the wake of landslips. The house to which the gardens belonged was converted into the Hotel Alexandra in 1901. The hotel continues to flourish and is an obvious place to stay for anyone researching the history of British ornithology.

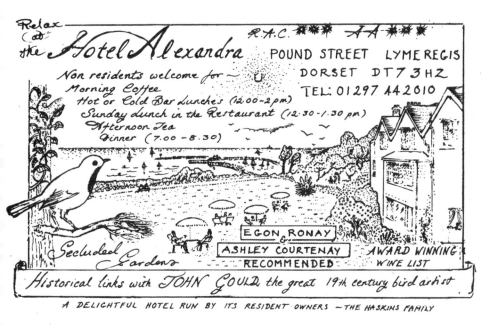

Relax at the **Hotel Alexandra** ... POUND STREET LYME REGIS
DORSET DT7 3HZ
TEL: 01297 442010

Non residents welcome for —
Morning Coffee
Hot or Cold Bar Lunches (12.00 - 2 pm)
Sunday Lunch in the Restaurant (12.30 - 1.30 pm)
Afternoon Tea
Dinner (7.00 - 8.30)

EGON RONAY
&
ASHLEY COURTENAY
RECOMMENDED

AWARD WINNING
WINE LIST

Secluded Gardens

Historical links with JOHN GOULD the great 19th century bird artist.

A DELIGHTFUL HOTEL RUN BY ITS RESIDENT OWNERS — THE HASKINS FAMILY

A recent advertisement of Lyme's Hotel Alexandra. Its original gardens were tended by the father of the great English ornithologist John Gould in the early 1800's. The hotel has been run by the Haskins family since 1982 and its association with ornithology is perpetuated by David Haskins who has built up a fine library of antiquarian bird books.

Within a decade after the death of John Gould another local figure to do with the history of ornithology puts in an appearance. Three miles to the west of Lyme Regis is a village called Rousdon to the south of which is a mansion containing a unique collection of stuffed British birds. The mansion was built in the 1870's by **Sir Henry Peek**. This Victorian gentleman employed the leading taxidermists of his day to provide him with specimens of all the 376 species included in the first list of British Birds drawn up by the British Ornitholgists' Union in 1883.

Only five species are missing from this collection, notably the **Great Auk** *Alca impennis* which, unknown to anyone at the time, had been persecuted to extinction. The collection includes a pair of **Passenger Pigeons** *Ectopistes migratorious*. Vast numbers of these birds blackened the skies of Audubon's America during their migratory activities and vagrants occasionally turned up in England. The last Passenger Pigeon died in Cincinati Zoo in 1914. The Rousdon collection is of outstanding importance not only to the ornithologist but also to the scholar of Victorian England.

The mansion and its grounds were taken over by Allhallows school in 1937 following the closure of its premises in Honiton which dated back to the sixteenth century. The school closed down down in 1998 and the Rousdon estate was sold off. The bird collection is still there, contained in its original cabinets which line the first floor of the mansion. Although now in private hands it is understood that there is a preservation order on the collection and its housing.

George Philip Rigney Pulman was another Victorian who had a passion for natural history. He loved fly-fishing along the River Axe, a site that we shall explore later on in our pursuit of birds. His experiences culminated in his famous "Book of the Axe" which contains a lot of information about the birds he saw. In 1857 he set up "Pulman's Weekly News". This local newspaper is still going strong.

The ornithological history of the twentieth century is overshadowed by the effects of modern methods of agriculture. The destruction of ancient hedgerows to make way for huge fields suitable for machines of mass cultivation, together with the spraying of herbicides and insecticides led to a decline in many species of birds that were once very much part of the English countryside. This decline is continuing today and the situation is not helped by the demand for housing. Another worry is the threat of oil spillage from tankers.

But it is not all bad news, at least around Lyme Regis. The six mile stretch of coast to the west of the town is a National Nature Reserve and much of the coast to the east is owned by the National Trust. Moreover, these areas are part of a Natural World Heritage Site called "The Jurassic Coast". With the exception of built-up areas it takes in the entire coastline between Exmouth in East Devon and Poole Harbour in Dorset, a distance of about 95 miles. (Specifically, between Orcombe Point and Old Harry Rocks).

This stretch of the English coast gained its World Heritage status in the year 2001, primarily in relation to its unique geology but also because of its wide range of habitats that attract some of our nation's rarest birds. Several of these birds will be considered later on, but first we should consider another aspect of ornithology that has a bearing on "The Jurassic Coast" – the evolution of birds.

THE JURASSIC CONNECTION

Lyme Regis is founded upon the muddy sediments of early Jurassic seas. These deposits were laid down about 200 millions of years ago and encapsulate the fossilized remains of innumerable sea-creatures. These remains are derived mostly from shellfish, notably ammonites and belemnites, creatures akin to the squids and octopuses of today. The fossilized bones of swimming reptiles that fed on these molluscs are buried with them together with bits of driftwood and the rare remains of dinosaurs that were washed into the sea.

The fossils are spewed out from cliffs onto the beaches around Lyme Regis and have made this town a mecca for the fossil hunter. For details see your author's "Fossil Hunting around Lyme Regis: A Practical Insight into the Jurassic Period " (ISBN 0-9520112-1-2). ISBN, by the way, stands for International System of Book Numbering. Any book that has been given an appropriate number is easy to locate.

The most cherished of all the fossils found around Lyme Regis are those of flying reptiles. Very few have been found. This is not surprising because the bones of these creatures, like those of birds, were full of air cavities as an adaptation for flight. But sufficient remains have been found throughout the world to picture early Jurassic skies teeming with flying reptiles. Birds have long been called feathered reptiles and following recent discoveries in China it is now generally accepted that birds either evolved directly from dinosaurs or shared a common ancestry with them.

The remains of the earliest undoubted birds are known only from Germany, in Jurassic deposits laid down long after those that now make up the cliffs of Lyme Regis. The first specimen was found in 1861 and consisted of the impression of a single feather. The bird to which it belonged was given the generic ("family") name *Archaeopteryx* from the Greek words for ancient feather. The impression was found on Jurassic limestone with an exceptionally fine grain.

This stone is used for the technique of printing known as lithography, explaining the full name given to the bird: *Archaeopteryx lithographica.* It is interesting to note that all of John Gould's prints were lithographic whereas those of Audubon were run off engraved copper plates. Hardly more than half a dozen individual remains of *Archaeopteryx* have turned up since 1861. They are probably the world's most valued fossils.

The Jurassic formations of England slope like a tilted pack of cards. The bottom of the "pack" makes up the base of the cliffs around Lyme Regis. Progressively younger Jurassic deposits hit the beaches as you go east. By the time you get to Kimmeridge, many miles away from the town and effectively 50 millions of years later, fossils contemporary with those of the Archaeopteran sediments of Germany are washed onto the beaches. But no one has discovered any trace of a fossilised bird along England's "Jurassic Coast" and should you find such remains around Lyme Regis then you will become world famous overnight and all the books on the evolution of birds would have to be rewritten!

The Jurassic sediments were overlain by sandy deposits of the Cretaceous Period. These give rise to the acidic soils that top the hills of the Dorset/ Devon coastal border. Heather grows on these soils to the liking of many rare birds, notably the Nightjar, a bird that we shall seek out.

Sensible Precautions

Most of the birds to be described here can be studied in safety from public footpaths. The cliffs around Lyme Regis are highly unstable however, and it important to heed all the warning notices that are stuck up along the coastal fringe.

Before bird-watching along the beaches it is essential to check tide conditions. There are numerous places where you can be cut off by rising seas. A major cut-off point is indicated in a map included here. But the coastal topography is always changing in detail in the wake of cliff falls and mud flows and it is advisable to seek local advice before setting out.

What to take

Many birds around Lyme Regis can be studied close at hand without the use of visual aids. These birds include not only common seaside birds such as the Herring Gull *Larus argentatus* but also rather rare species such as the Rock Pipit *Anthus spinoletta* and the Purple Sandpiper *Calidrus maritima.* But to get the best out of your bird-watching you will need a pair of **binoculars**. This can be something of a major expenditure but is normally a sound investment. There is a big market for this "tool-of-the-trade". You are spoilt for choice and deciding what to buy can be a bit of a headache.

If you do not already own a pair of binoculars then seek out the advice of the experienced bird-watchers you will certainly meet as you explore the various sites of ornithological interest described in this booklet. These people are often pleased to let you try out their equipment and show you what birds are about. You should find the local experts that utilise the hide on the Axe Estuary especially helpful. If you are new to "serious" bird-watching then you might well be surprised how quickly your knowledge builds up in the company of friendly experts. You will certainly gain an insight into ornithology that is difficult to acquire from reading alone.

Most bird-watchers carry a **Field Guide** to birds, a book geared up to recognition of species and containing clear-cut annotated illustrations prepared by professional artists. There are many of these guides, perhaps too many, and once again you are spoilt for choice. A work that has stood the test of time is "A Field Guide to the Birds of Britain and Europe" by Peterson, Mountfort and Hollam, published by Collins.

What to expect

Many species of birds live out their lives in and around Lyme Regis and these **residents** are always evident. Herring Gulls and Rock Pipits are typical examples considered here. Other birds that we shall take in are seen only at certain times of the year. These include **summer visitors** that come here only to breed (very often from Africa) such as the Nightjar; **winter visitors** that stay here after breeding elsewhere, such as the Purple Sandpiper which raises its chicks within the Arctic Circle;

passage migrants , birds that utilise Lyme Regis as a sort of stepping stone, feeding *en route* before heading inland to nest such as the Wheatear; and ***vagrants***, birds that have strayed a long way from their normal homes and which make up the bulk of species that excite the "twitcher". This is the the nickname given to the bird-watcher who seeks out rare birds and is a term used somewhat tongue-in-cheek to distinguish this kind of ornithologist from the "bird-lover" who gets his enjoyment from every bird he sees.

The categories listed above are by no means exhaustive and ignore the movements of populations within species. It is also worth noting that many birds are more often heard than seen, especially during the breeding season when they are hidden by foliage.

Excellent recordings of the songs and call notes of British birds are available and the more you listen to them the better. It is notoriously difficult for most people to remember, let alone mimic, the songs of even our commonest birds. Some people have an exceptionally "good ear", sufficient to enable them to tell apart the individual voices that make up a dawn chorus. If you have this rare gift then you will soon find that your company is much sought after by fellow ornithologists!

Code of Conduct

There are strict laws in relation to the protection of British species of birds. These legal aspects hardly concern the bird lover who will do his best to avoid disturbing the creatures the admires. But it is all to easy to forget to close a gate, letting loose sheep and cattle to the infuriation of local farmers. Like everyone else who explores the English countryside the bird-watcher is expected to abide by "The Country Code".

What to wear

Clothing that blends in with the countryside is the rule unless your bird-watching is restricted to species such as the Herring Gull that avidly seeks your company in the pursuit of food. The area covered here is renowned for winter visitors, notably the Purple Sandpipers of Lyme Regis and the thousands of waterfowl that hone into the Axe Estuary. Although you may need to wrap up warm when getting out and about to see these birds, many of them can be observed in the comfort of a car parked along the banks of the estuary. It is is also worthy of note that the bird hide of the Axe Estuary is roofed over and is essentially an all-weather facility.

Where to stay and where to go

The map which follows is a rough plan of local sites of ornithological interest and is intended to whet your appetite before we consider them in detail. It will be evident that either Lyme Regis or Seaton is an appropriate base for the ornithologist. There is far more accommodation in Lyme Regis for the visitor than in Seaton and there is no denying that Lyme Regis is the more attractive of these towns. But if you are a thinking of retiring to the area then Seaton is a better bet if only because of the outstanding facilities of the Axe estuary. Assuming that you are visitor staying in Lyme Regis you will have no difficulty in watching the first bird on our agenda. Rather, this ubiquitous bird will be watching *you* from the moment of your arrival!

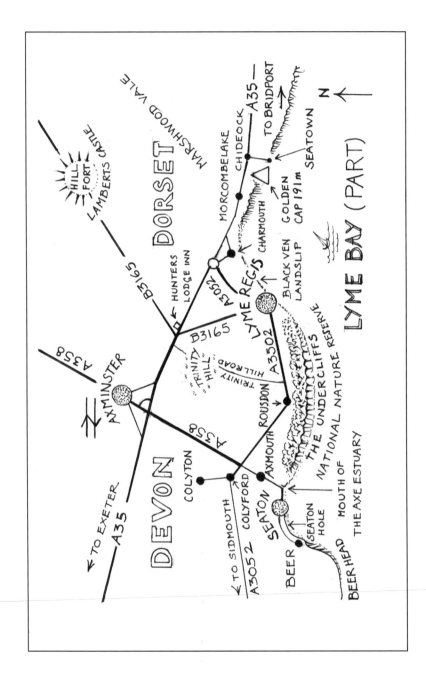

Road connections (not to scale) between various sites of ornithological interest around the Dorset / Devon coastal border

BIRDS OF LYME REGIS

The Herring Gull

The Herring Gull *Larus argentatus* is very much a resident of Lyme Regis where it breeds on roof-tops, believing them to be the ledges of a cliff – its natural habitat. Its nest is an untidy assemblage of grass, moss, and twigs. Its favourite nesting site appears to be the crevice between a chimney stack and a pitched roof. Debris falling from the nest clogs up guttering making the bird a nuisance to the householder. It is careless in disposing of its natural waste to the annoyance of the motorist who parks close to buildings and finds his car splattered with white droppings. The bird is also notorious for ripping open the plastic bags used for putting out rubbish and which have replaced "gull-proof" dustbins.

Along the seafront you will see notices put up by West Dorset District Council asking you not to encourage gulls into the town by feeding them. Meanwhile the birds go hungry and their plight has not been helped by a bad Press. One visitor to Lyme Regis was apparently hit on the head by a whelk shell dropped by a Herring Gull – an off-beat example of the bird's well known habit of attempting to crush open shellfish for food. Other visitors have certainly had their fish and chips unceremoniously plucked from their hands. Even your author's head has been targeted by a Herring Gull in a wasted attempt to secure nest material (I am bald on top but my whiskers are flourishing). Every now and again you hear about a tourist who has actually been attacked by the bird, notably a tourist visiting Cornwall, the county of residence of the late Daphne du Maurier whose short story 'The Birds' took avian dominance to the limit and was made into a scary film by Alfred Hitchcock.

But no seaside town would be the same without its seagulls and you would be hard pushed to find scientific evidence that these birds are a serious threat to your health and safety. Moreover, the proximity of the Herring Gull enables the ornithologist to study its habits in detail. Much of our understanding of bird behaviour (and of behavioural mechanisms in general) followed research on this species by the Dutch scientist **Niko Tinbergen**. His book "The Herring Gull's World" is a classic of avian literature and a joy to read. It was published by Collins in 1953 as part of the famous New Naturalist Series.

One of Tinbergen's discoveries concerned the function of the red spot that tips the beak of the adult gull. He showed that the spot aroused an innate response in the hatchling to peck it and a consequent urge in the parent to cough up regurgitated food. He also demonstrated that such "sign stimuli" play a vital part in the reproductive life of many other creatures. All this research resulted in a fundamental concept of instinctive behaviour, that of **releaser mechanisms.**

Another legendary ornithologist of the 20th century is **Konrad Lorenz**. He is especially famous for his studies on **imprinting**, the term used to describe the way in which a developing bird or mammal normally acquires an indelible image of its parent. An endearing image in the history of ornithology is of Konrad Lonrenz with a batch of goslings following his footsteps. He had succeeded in imprinting the chicks with his own image. Tinbergen and Lorenz worked together on many aspects of bird behaviour. Their efforts were rewarded by a Nobel Prize in 1973. They shared it with Karl Von Frisch, famous for his research on bees.

As you walk around Lyme Regis you will soon be asking the same questions that Tinbergen put his mind to. How, for example, do Herring Gulls tell each other apart? Certainly, the male is slightly bigger than the female, but otherwise they look very much alike apart from variations in the black and white markings on their wing tips.

And what are the meanings of the various calls that the birds make ? The birds are especially vocal during the breeding season. If you hear something like a repetative laugh coming from a roof top then look up and you will probably see a male gull mounting a female! Other vocalisations of the Herring Gull include a "gaa-gaa-gaa" at the approach of a stranger, hissing by a cowering chick begging for food, and deafening screeches when a pair of gulls meet or are thrown food and which are often accompanied by up and down movements of the head and neck.

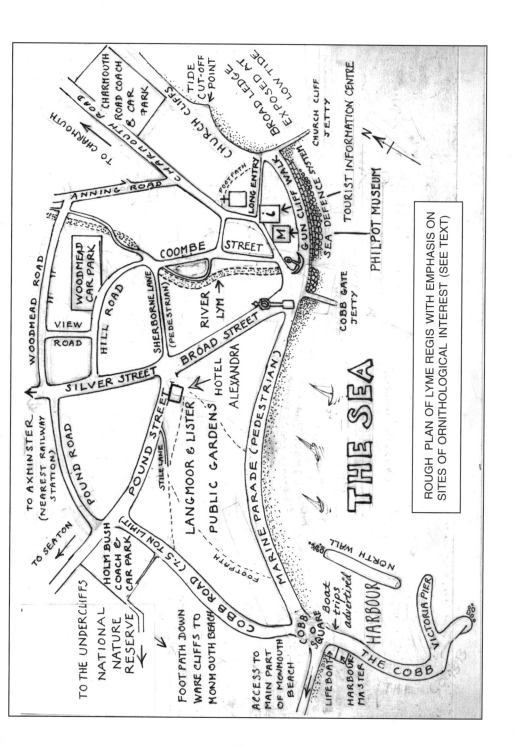

ROUGH PLAN OF LYME REGIS WITH EMPHASIS ON SITES OF ORNITHOLOGICAL INTEREST (SEE TEXT)

The Herring Gull takes up to four years to attain sexual maturity. It begins life as a downy chick with a somewhat grotesque beak which appears out of proportion to the size of its body. The down is greyish-white mottled with dark spots and blends in remarkably well with its surroundings, cliff-face or roof-top. In contrast, the mouth of the chick is bright pink and this colour acts as a trigger for the feeding parent.

This feature – a colourful gape evoking parental feeding behaviour – is common to many birds. The mouth of the fledgling Blackbird, for example, is a vivid orange. And in many species of birds the mouth is a riot of colourful markings as in the **Bearded Tit** *Panurus biarmicus.* You will not need to disturb nests in order to appreciate them – fine illustrations by Philip Burton feature in Colin Harrison's book "A field Guide to Nests, Eggs, and Nestlings of British and European Birds", published by Collins (ISBN: 0-00-219249-7). The significance of these patterns is especially evident in those species that lay their eggs in the nests of other birds. The mouth markings of the hatching "parasitic bird" exactly mimic those of the young of its host. The **Common Cuckoo** *Cuculus canorus* is a classic example. The colours and markings of the eggs are likewise replicated.

The Herring Gull chick grows rapidly and within about seven weeks after hatching it is as big as the adult and ready to fly. It feathers are mostly dark brown. By the end of the year the youngster is in serious competition with adults for food. It still continues to beg, adopting the cowering posture and hissing, but its parents will have nothing to do with it and chase it off. This remarkable switch from parental care to bullying is a demonstration of the way in the lives of birds are governed primarily by instinctive reactions.

The youngster moults during its second year and its brown feathers are mostly replaced by white ones. It then looks something like the adult but it is easy to recognise as an an immature gull from the brownish and grey patches on its wings. In the spring of its third year the bird moults again, adopting the plumage of maturity. The pristine plumage of the breeding Herring Gull gives way to a rather tattered investment, an aspect enhanced by the flecks of grey feathers that develop around its head and neck as winter approaches.

Fancy a chip? Gulls are ravenous during the winter. At this time the neck of the Herring Gull is flecked with grey streaks. The bird competes with visiting Black-headed Gulls whose heads have lost the black hoods of their breeding plumage apart from a dark smudge behind the eye.

In order to study the nesting habits of the Herring Gull, you will need to find a room in a hotel overlooking a rooftop. Such a viewpoint is rare bearing in mind the nest site is always noisy and that no hotelier wants to encourage the bird to breed when his guests expect a good night's sleep!

In contrast, you can hardly avoid noticing the feeding habits of this ever-hungry bird about the town. Especially fascinating is the way in which it often shuffles its feet on the slopes of the public gardens – much to the amazement of visitors. It is said that the gull is imitating the pitter-patter of rainwater in an attempt to bring earthworms to the surface so that it can eat them. They do the same sort of "dancing" on sandy patches on the sea-shore, presumably in pursuit of marine worms. Just as curious is the way in which the gull pokes around the footpaths of the gardens during the summer, licking up ants.

These antics together with "shell dropping" indicate that the Herring Gull is much more than just a scavenger. But when food is plentiful it wastes no time in gobbling it up. If you throw a loaf of bread at the bird it will disappear in a matter of seconds. The food is immediately stored in its expandable gullet and digested after the bird has flown away to safety.

The Herring Gull is an opportunistic feeder that exploits our untidy habits but otherwise it is very much a seabird. This is evident in the structure of its skull as illustrated below. Above the eye sockets are sausage-shaped depressions which, in life, accommodate glands that pump out salt. These organs are common to seabirds and enable them to drink sea water with impunity. This is something that we and most other mammals can't do: our kidneys cannot expel the salt fast enough to avoid the building up of a lethal concentration within the blood.

The skull of a Crow (left) is smooth on top whereas that of the Herring Gull (right) has depressions above the eye-sockets. In life, these cavities contained massive glands for pumping out salt from the blood. The illustration is based upon specimens found at Deadman's Cove situated between Seaton and Eype Mouth.

OTHER GULLS & SIMILAR BIRDS

Whereas the Herring Gull completes its lifecycle on the roof tops of Lyme Regis, the **Black- headed Gull** *Larus ridibundus* breeds at sea-level and a long way from the town, such as on Brownsea Island off Poole harbour, thirty miles to the east. But after the breeding season, adults and their progeny fan out out along the Jurassic Coast. Many of these spend the autumn and early winter in Lyme Regis where they compete with the Herring Gulls for food. By this time most of these visitors have lost the black feathers around their heads apart form a few that make up a dark spot behind the eye.

This gull is roughly two thirds the size of the Herring Gull. Its crimson bill and legs are notable features and its grey-backed wings have a white leading edge. The tips of its wings sport a pattern of black and white "mirrors" as in the Herring Gull. Its calls are more highly pitched than its bigger relative, scratchy rather than raucous. It keeps its distance but it will go for any food you throw up in the air, taking it on the wing.

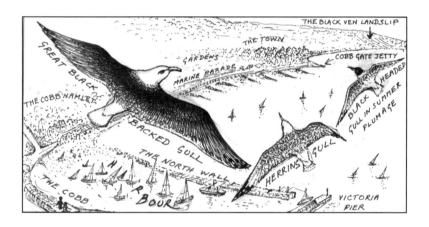

The **Great Black-backed Gull** *Larus marinus* is the largest gull in the Northern Hemisphere.When seen at a distance roosting with Herring Gulls, as on Broad Ledge, they don't look that much bigger in comparison. It is only when the bird takes to the air and its black wings are expanded that its size is readily appreciated. Its flight is more leisurely than that of the Herring Gull.

One or two pairs are always to be seen around Lyme Regis but you are unlikely to get close to them. Unlike the gulls that we have considered they instinctively avoid our company and never "beg" for food. They are more at home on off-shore islands where they are at the top of a food chain, stealing food from other birds. They sup-plement their diets with live food, notably the chicks of other gulls. They also take small mammals, including rats.

The best place to study this species is around the Axe Estuary where large flocks overwinter. This site is also commandeered by numbers of its smaller relative, the **Lesser Black-backed Gull** *Larus fuscus.*

The first and second year youngsters of many species of gulls are mostly brown and can not be told apart without a lot of experience. And the adults of some species are almost identical apart from certain details. The **Common Gull** *Larus canus*, for example, is the same size as the Black-headed Gull and looks just like it during the winter except for its greenish bill and feet. The bird does not live up to its name locally but one or more individuals are sometimes seen amongst a flock of Black-headed Gulls. Keen bird-watchers often hone in with their binoculars to a flock of roosting gulls in the hope of seeing unusual species that have hitched a lift as it were. These species have everything to gain when it comes to security. Watch any bird feeding in the open. Half the time it will looking out for danger. The more birds feeding together the greater is the chance that one of them will spot a potential enemy in good time to raise a general alert.

Terns are sometimes mistaken for gulls but they are much more streamlined. They have aptly been described as the swallows of the sea. These beautiful birds work hard for a living, plunge-diving into shoals of fish, often from a great height.

All terns breed in colonies. One of these is situated on Chesil Beach just beyond Abbotsbury and about twenty miles to the east of Lyme Regis. The eggs are laid on shingle, without nest material, but blend in with the pebbles. The birds can not survive in the wake of human disturbance. It is for this reason that the entire colony is cordoned off during the spring and early summer. Later on many birds from this and other colonies fish along the Jurassic Coast. They often come very close inshore, a few meters away from a sunbather in Lyme Regis, but are more often seen further out, giving themselves away by their high pitched calls. The species that you are most likely to see is the **Sandwich Tern** *Sterna Sandvicensis.*

Feeding habits alone are sufficient to tell the scavenging Herring Gull (bottom left) from the aerobatic tern, which plunge-dives for fish.

CORMORANTS

The distinctive silhouettes of these dark fish-eating birds show up well against the background of the open sea. When seen flying at a distance they look like black matchsticks with wings. They follow a straight course a metre or two above the waves as if they knew exactly where they were heading. When fishing, the Cormorant (*Phalacrocorax carbo*) dives from a sitting position at the surface. It often roosts with wings outstretched as if to dry them. But it is also possible that this characteristic posture has a lot to do with diet. The Cormorant swallows its prey alive and folded wings might get in the way of an expanding breast cage. The bird is adept at catching eels but it struggles to engulf them. If you see a Cormorant with a white patch on its thighs then it will be in breeding condition. Cormorants are always to be seen at low tide roosting on the seaward tip of Broad Ledge in the company of gulls.

PURPLE SANDPIPERS & OTHER WADING BIRDS

Purple Sandpipers *Calidris maritima* breed within the Arctic Circle during our English spring and summer. They then fly south and about a dozen birds of this species overwinter around Lyme Regis. The flock normally keeps together, probing around Broad Ledge when the tide is out and roosting about the Cobb when the tide is in.

The bird is no bigger than a starling. Its feet are yellow. So too is the base of its beak which fades out to a black tip. Its overall plumage is mottled with shades of brown and dark grey except for whitish underparts. The overwintering birds of Lyme Regis hardly live up to the common name for the species.

The birds blend in remarkably well with rocks and seaweeds. The flock can be difficult to locate but once you have found it you can study the birds closely. They are tame in the sense that they will carry on feeding until you get within a couple of metres of them. Get any closer and they will hurry along in single file before suddenly taking to the air in a V-shaped formation that keeps low down and disappears behind your back. The flock then immediately resumes feeding and it may take several minutes to relocate.

The precise location of the roosting quarters of the birds vary according to weather conditions. When it is fine they rest on the boulders at the end of the Cobb as illustrated above. When the waves are bashing these stones they sit tight within the crevices of the southern face of Victoria Pier. To make them out you will then need to look directly downwards from the walkway behind the buildings on the pier. Look for thin beaks poking out of a wall!

The Purple Sandpiper is a typical wading bird, spending much of its day with its feet awash. But it is by no means common and some twitchers visit Lyme Regis simply to tick it off.

In contrast, the **Oyster Catcher** *Haematopus ostralegus* is a familiar British wader and can be seen at any time of the year on the shores of Lyme Regis. This shy bird is the size of a small gull and is unmistakable because of its colourful appearance and distinctive silhouette. Its rather heavy orange bill is straight and long. Its legs are pink.These features contrast with its black shoulders and white underparts. In flight the white bars on the upper surfaces of its wings are distinctive. It is far less approachable than the Purple Sandpiper and will normally fly off as soon as it sees you, uttering its alarm call, a series of high-pitched peeps.

Several other common waders explore the ledges around Lyme Regis. These are mostly small visitors about the same size as the Purple Sandpiper and often smaller. They keep their distance and are difficult to make out. The best time to look for them is during the autumn when gales wash up masses of weed and flotsam on to the seashore, especially where this debris piles up against the Cobb in keeping with prevailing south-westerly winds.

The decaying weed provides food and shelter for countless numbers of tiny creatures that the birds seek out. These creatures include sandhoppers, hunchbacked crustacea (members of the crab family) no longer than a finger nail and flattened from side to side and which hop out of sight when uncovered, and tiny flies and their larvae. Birds unrelated to waders join the feast, notably Rock Pipits and Wagtails.

All these birds blend in with their backgrounds and although scores may be present they are seldom noticed until they are disturbed and take to the air. Many of them have brownish mottled backs that render them almost invisible when they keep still.

Others are white all over apart from a pattern of black shapes. An excellent example of a bird of this sort is **Ringed Plover** *Charadrius hiaticula*. When it is viewed against a grey background, as on a mudflat, the black ring that skirts its breast and other streaks that adorn its head are very evident. But when the bird is foraging about shingle these markings conform to the irregular shapes of pebbles and their shadows. Such a phenomenon is known as **Disruptive Camouflage,** well known to experts of human warfare. Other avian examples include the terns that nest on Chesil Beach. Nearer to home, the Pied Wagtail is always difficult see when it forages about Monmouth Beach.

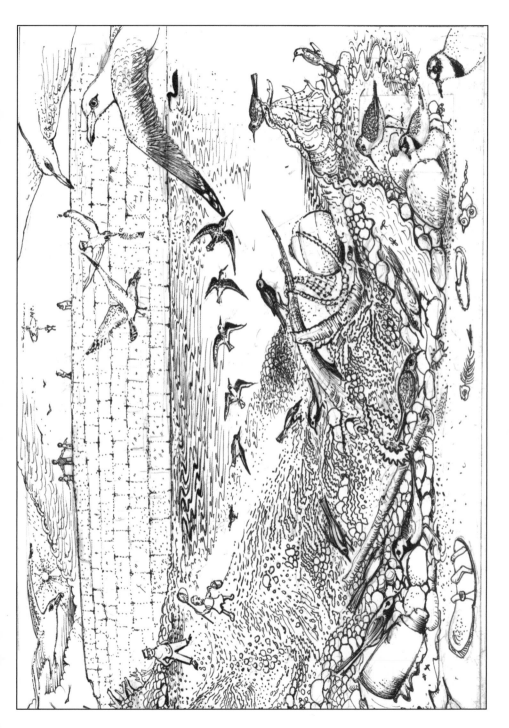

ROCK PIPITS & WAGTAILS

Lyme Regis has a thriving population of Rock Pipits. They are about the same size as sparrows and look very much like them from a distance. For this reason most visitors to the town rarely give them a second glance. But whereas the familiar House Sparrow *Passer domesticus* has a stout beak for crushing seeds, the Rock Pipit *Anthus petrosus* has a pointed beak for capturing flies and other tiny creatures.The beak of the Dunnock *Prunella modularis* (commonly known as the "hedge sparrow") is also thin but this bird lacks the streaks that adorn the breast of the pipit.

The Rock Pipit lives up to its full name, flitting about rocks and frequently uttering its call note: " pee-pip pee-pip." Not that the beaches are particularly rocky around Lyme Regis; it is the nature of cliffs that suits them. The base of the cliffs is made up of alternating layers of tough limestone and weak shale, a Jurassic formation known as the Blue Lias. Waves bash the cliffs and rainwater drips down from above, giving rise to innumerable crevices which are commandeered by breeding colonies of the small invertebrates upon which the pipits feed.

The larger crevices provide nesting sites for the bird. Bearing in mind the instability of the cliffs, the survival of the Rock Pipit is something of a local curiosity.The bird is certainly something of a desperate chancer. In 1997 the abandoned nest of a Rock Pipit containing five eggs was found in a hat left in the bow of a boat moored in the harbour.

There is undoubtedly a lot of competition for the few holes in the old and high walls that form a rising backcloth to the sea-defence system on the eastern fringe of the town. A pair of Rock Pipits usually nests in one of the holes every year, well above the heads of holiday-makers. This part of the the town is an excellent place to study the bird where it is often seen, either running along the pavement or perching on the low seawalls. As soon as you get to within six feet of them they instantly fly away however.

In contrast to the Rock Pipit, the **Meadow Pipit** *Anthus pratensis* is a common resident bird of open countryside throughout England. During the summer another pipit, the **Tree Pipit** *Anthus trivialis*, visits our islands to breed, notably on heathland.

All pipits look similar and if you can identify one species from another at a glance then consider yourself a top ornithologist! The location of these birds - rocks - meadows - heathland - are the first clue. Then comes confirmation under binoculars, honing into such details as the colouration of the birds' outer tail feathers which vary from white to pale grey according to species. Pipit enthusiasts are always on the look out for the **Water Pipit** *Anthus spinolleta* formerly considered to be a sub-species of the Rock Pipit. The debate is still going on.

The tail feathers of pipits are rather long but this feature is especially noticeable in their very close relatives: wagtails. Another feature that links these birds together are their exceptionally long hind claws.

In contrast to the various species of pipits, our two resident wagtails – the **Pied Wagtail** *Motacilla alba* and the **Grey Wagtail** *Montacilla cinerea* – are very easy to tell apart. The Pied Wagtail is black and white. The Grey Wagtail is mostly yellow. Both species are very common birds that breed around Lyme Regis.The Grey Wagtail gets its name from the greyish parts of its plumage but is otherwise a distinctly yellow bird. It should not be confused with the **Yellow Wagtail** *Motacilla flava*, a rather uncommon summer visitor.

As we have noted, Rock Pipits and Pied Wagtails are often seen together foraging for food on Monmouth Beach, especially after a lot of seaweed has been washed up. When food is short on the beaches during the winter both species are often seen running along Broad Street in search of grubs between cracks in the pavement. They are sometimes joined by Grey Wagtails but these birds are more often seen along the River Lym (also spelt "Lim") where they breed. You are almost certainly to see a pair of them as you make your way out of town up the river in search of a rather rare bird: the **Dipper**.

DIPPERS

The **Dipper** *Cinclus cinclus* gets its name from the way in which its body bobs up and down when the bird is standing, as on the boulder illustrated above. Its shape is similar to that of a wren but the Dipper is bigger, darker, and sports a white apron. It seeks out the tiny invertebrates that flourish in the well aerated waters of fast moving rivers and streams. To catch these creatures the Dipper literally walks under water, but the bird is also adept at picking up insects that fly about its habitat.

This remarkable species breeds along the river Lym. The river is served by five tributaries, each draining from springs that arise at the heads of valleys that skirt Lyme Regis to the north. These tributaries pass through soft sandy deposits and are hardly more than meandering trickles. But they all eventually meet up and their combined strength has carved a narrow deep-sided gully through the tough Jurassic limestone that underlies the town at sea-level. The waters pour into the sea under a narrow bridge at the foot of the town.

The bird flits about the river Lym close to town but is difficult to see because of the dense vegetation that clothes the river banks. The species is very much a local talking point ("have you seen the Dippers lately?") and there is nothing to be lost in asking local residents about the latest sightings as you follow the river out of town.

The Dipper is uncommon and there are probably not more that about 30 pairs in the whole of Dorset (see "The Birds of Dorset" published by Christopher Helm in 2004: ISBN 0-7136-6934-9).

Even if you don't get to see the bird in Lyme Regis, a Dipper hunt up the River Lym is always worthwhile. Grey wagtails are common and as you make your way out of town you will soon be in good Buzzard country.

BUZZARDS

The **Common Buzzard** *Buteo buteo* lives up to its name on the Dorset/ Devon Coastal border but this huge bird of prey keeps mostly out of sight. It spends most of the day perching motionless in tall trees or on shrubby branches sprouting out of from the top of inaccessible cliffs. During the summer it is hidden by leaves and during the winter its mottled feathers of brown and grey blend in perfectly with branches and twigs. It sometimes roosts on telegraph poles close to a road but otherwise the bird is difficult to locate unless it is moving. It is only when it soars high up in the sky that it is noticed by a visitor to the area with only a fleeting interest in birds.

The Buzzard is never in the air for very long however. As soon as it spots its favourite prey – a rabbit – it plunges down to earth.There are thousands of rabbits around Lyme Regis in the soft sand that tops the cliffs and the hillsides and in which the rabbits easily burrow. And should the Buzzard linger in the air it is soon molested by a variety of unrelated birds that have an instinctive urge to attack it, notably crows and gulls, but smaller birds often join in. The wings never appear to touch but sooner or later the aggravated Buzzard returns to its perch. The reason for this aggressive behaviour is by no means clear. Buzzards rarely feed on the eggs and chicks of its molesters although other birds of prey are notorious for making a meal of them. And it is curious that the fish-hunting Heron is similarly attacked.

STONE CHATS

The **Stonechat** *Saxicola torquater* is sometimes mistaken for a robin when seen from a distance. But in contrast to the "friendly" British Robin, the Stonechat avoids human company and is very much a "wild" bird. It is common all around the Dorset/Devon coastal border where it spends most of its time flitting about the branches of low shrubs beneath which it makes its nest. It often perches on a branch sticking out of the top of a bush and is sometimes seen alighting on a tall flower which bends over with its weight. The bird is a strong flyer and can show up on the top of any building in the area and when you are least expecting it.

A convenient place to look for the Stonechat is along Monmouth Beach. As indicated in the illustration it often puts in a seemingly proud appearance on the tops of the bushes that cover the sloping site of a very old landslip. The pebbles of the beach serve as a reminder of how the bird got its common name – its alarm call sounds just like stones hitting together.

The birds are as much at home on a heathland as they are on rough ground close to the sea so long as there are bushes about for breeding purposes. We shall meet this attractive species again when we take in the heather-covered hills to the north and east of Lyme Regis in pursuit of Nightjars and Dartford Warblers.

The male Stonechat is especially distinctive. It sports a chestnut-red breast and a black hood. The female is pinkish all over and all the more liable to be mistaken for a Robin.

WHEATEARS

The Stonechat breeds all over Europe and resides largely within this land mass. In contrast, its close relative the **Wheatear** *Oenanthe oenanthe* is a trans-continental species, overwintering in equatorial Africa and breeding in the northern hemisphere. It is the earliest of the spring migrants to reach our shores, showing up as early as the first week of March and well before much more familiar migrants such as the Housemartin, Swallow and Swift.

Like many migrating birds crossing the English Channel, Wheatears utilise the Isle of Portland as a sort of stepping stone before fanning out along the Jurassic Coast. The Islanders once captured Wheatears by the thousand for shipment to the Victorian dining tables of London.

A few Wheatears invariably show up in Lyme Regis where they are sometimes seen flitting about the boulders of its eastern sea-defence system and the big stones that litter Monmouth Beach. The bird sticks close to ground and when viewed from above its white rump is very evident together with a black letter "T" on its tail. Not so easy to make out is the reddish-orange tinge of its breast feathers.

The bird soon heads further north or west in search of its breeding habitat – typically rough ground associated with moorland. Many Wheatears end up as far away as Canada. Various sub-species of this slick little bird are recognised of which the "Greenland Wheatear" is the best known. It arrives later on our shores and is noticeably larger than the first arrivals. But whatever its type and destination the Wheatear always returns to Africa, a journey that may take in over two thousand miles. No other "European" bird of similar size can match the Wheatear as a migratory species.

FURTHER AFIELD

RARE COASTAL BIRDS: RAVENS & PEREGRINE FALCONS

The birds that we have so far considered can be seen within or very close to Lyme Regis. Three of these species are rather scarce : the Rock Pipit, the Purple Sandpiper and the Dipper. In contrast, you will need to go further afield to see a Raven or a Peregrine Falcon. These species are very rare nationally and the Jurassic Coast is one their strongholds. They fly about the cliffs to the east and west of Lyme Regis but are seldom seen close to the town.

The **Raven** *Corvus corax* might be mistaken for its smaller and common relative the **Carrion Crow** *Corvus corone.* Distinctive features of the Raven include its massive beak, ragged neck and wedge-shaped tail. But these details are not easy to make out when the birds are twisting and turning high up in the air. A better visual clue might be the more leisurely flight of the Raven but there is one aspect that gives this bird away – its voice. This is a deep croak which you can imitate by pinching your nose and saying, very slowly, "Pruck". Like the Buzzard it is mobbed by other species, including the Carrion Crow, as indicated in the bottom right of the illustration above.

The Raven has a huge territory and it is a waste of time scanning the sky in the hope of seeing it and it is far better to keep your ears pinned back. The bird can be studied at close quarters where it makes its huge nest of twigs but locally it chooses an inaccessible site which is often shrouded by trees.

The distinctive face of the **Peregrine Falcon** *Falco peregrinus* is featured on the the cover of this book. This beautiful raptor is said to be the fastest bird in the world,clocking up over 200 miles per hour in pursuit of its principal target – pigeons. It has a history of persecution and it is only within recent years that this species has become re-established as a nesting species on the Jurassic Coast. During the Second World War it was shot on sight in case it took pigeons carrying messages across the English Channel. After the war the bird was an indirect victim of pesti-cides.

The Peregrine has been known to pluck a pigeon from a pavement in the heart of Lyme Regis. But such an event is exceptional and in order to be fairly certain of see-ing the bird you will need to view the cliffs where it breeds. If you are prepared for a lengthy walk over shingle then hone your binoculars beneath "Twin Peaks", about a mile to the east of Charmouth.

Twin peaks and twin pursuits. Crumbling Jurassic cliffs spew out the remains of ancient sea creatures on to the beaches to the joy of the fossil hunter who always look down. At the same time pinnacles provide precarious nesting sites for Peregrine Falcons to the joy of the bird-watcher who is obliged to look up.

Like the Raven, the Peregrine roams far and both species can show up anywhere along the coast. A convenient place to make its casual acquaintance is around the cliffs of Beer Head in Devon. This promontory is very much on our itinerary but before exploring the cliffs to west of Lyme Regis we shall make a detour inland in search of a remarkable bird that hunts at night and which is almost as agile as the Peregrine.

TRINITY HILL: NIGHTJARS & OTHER HEATHLAND BIRDS

The **Nightjar** *Caprimulgus europaeus* is one of several birds that are dependent upon heathland for their survival. Most of the hills to the north of Lyme Regis were once covered in heather with spatterings of gorse. Most of this has disappeared under intensive agriculture, conifer plantations and housing development. A precious remnant of heathland tops Trinity Hill, four miles north-west of Lyme Regis. It might well have disappeared if it were not for local hero Ron Alford who insisted on his grazing rights.

The lowland heath of Southern England is essentially man-made. Without light grazing the heather is soon overtaken by a larger plants such as bracken, culminating in woodland. For centuries vast areas of heathland in Devon and Dorset were unwittingly retained because of the common rights of grazing. The nibbling activities of sheep and cattle ensured that bracken and gorse never ousted the heather. This plant burns easily when dry but its roots survive and heathland fires naturally contribute towards its survival by killing off its floral competitors and providing light for its fresh shoots. It was common practice to set fire to parts of heathland to ensure the survival of heather.

A lot of England's heathland was enclosed and "improved" during the middle of the nineteenth century. By the late 1800's the Dorset novelist Thomas Hardy was complaining about the loss of parts of Eggardon Heath which features so much in his work. Little of this heath is left and "Thomas Hardy Country" is something of a misnomer.

The decimation of England's heathland is now a matter of international concern. To get to grips with the problem and learn about the wildlife associated with this habitat consult David Allen's delightful "Heathland in East Devon and the Blackdown Hills", a beautifully illustrated eighty-page booklet published in 2004 by the Blackdown Hills Rural Partnership.

Trinity Hill was saved as a Local Nature Reserve in 1995. It is especially renowned for its Nightjars. They hone into the hill to breed at about the end of April after leaving their wintering quarters in Africa. During the day they lie low about the heath and are very difficult to see. They sit tight amongst the heather or on an old log, blending in perfectly with their surroundings. If you disturb one of these birds then it will take off silently, looking like a cut-out figure of its terrain, not unlike the alien featured in "Predator", a thrilling 1987 film directed by John Tierman and featuring Arnold Schwarzeneggar.

As dusk approaches the Nightjar begins its incessant churring that gives rise to the second part its name. This is a monotonous and somewhat eerie call which is never forgotten once heard. If you are scared of the dark then join one of the Nightjar forays in June organised by local Conservation authorities (Contact the Seaton Tourist information Centre for details). Bring a torch! The mouth the Nightjar is fringed with hairs to entrap its prey, typically moths.

Many interesting birds can be seen on Trinity Hill during the day, including Stonechats and Meadow Pipits, birds that we have already considered. Rare summer visitors to the Hill include the **Hobby** *Falco subbuteo* which looks exactly like a miniature Peregrine Falcon. It goes for big insects and small birds. The days have long gone when **Shrikes** were common about the hill. They are often called "butcher birds" because of their habit of impaling food (caterpillars and the like) on thorny branches of gorse and other shrubs which function as a sort of larder.

As indicated in the sketch below much of the Reserve is surrounded by coniferous plantations. These are always dark places. The evergreen trees shut out light and pine needles take a long time to rot away, halting the development of a rich undergrowth of flowers and shrubs. That said, the plantations are favoured by a variety of **Titmice** and the **Goldcrest** *Regulus Regulus*, Britain's smallest bird. All these birds are difficult to make out in their dark environment but if you hear what sounds like an incessant toy sewing machine then it will almost certainly be the voice of a Goldcrest.

As the trees are felled for timber, the brush-littered and sunlit ground provides a habitat for rapidly breeding insects, much to the benefit of Nightjars. It follows that the favoured breeding sites of these curious birds vary about Trinity Hill from one year to another.

The ground gets soggy and adders are common. You are advised to wear wellington boots. If a dog is with you then it would be unkind to let it off its lead, especially when birds are breeding about the heather.

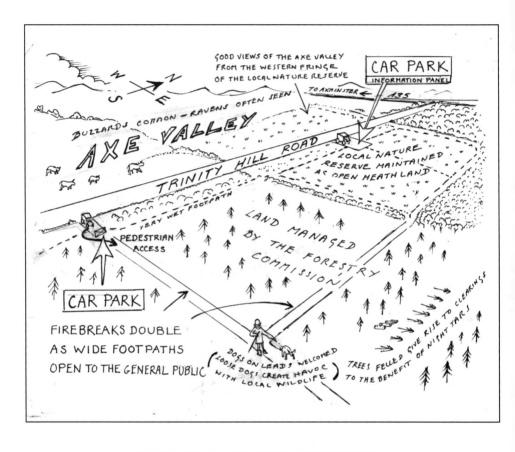

The following text labels appear within the sketch:

GOOD VIEWS OF THE AXE VALLEY FROM THE WESTERN FRINGE OF THE LOCAL NATURE RESERVE

CAR PARK

INFORMATION PANEL

TO AXMINSTER ← A35

BUZZARDS COMMON — RAVENS OFTEN SEEN

AXE VALLEY

TRINITY HILL ROAD

LOCAL NATURE RESERVE MAINTAINED AS OPEN HEATH LAND

VERY WET FOOTPATH

PEDESTRIAN ACCESS

LAND MANAGED BY THE FORESTRY COMMISSION

CAR PARK

FIREBREAKS DOUBLE AS WIDE FOOTPATHS OPEN TO THE GENERAL PUBLIC

DOGS ON LEADS WELCOMED LOOSE DOGS CREATE HAVOC WITH LOCAL WILDLIFE

TREES FELLED GIVE RISE TO CLEARINGS TO THE BENEFIT OF NIGHTJARS

ROUGH "BIRD'S EYE" SKETCH OF TRINITY HILL

The hill is an important site for breeding Nightjars. It was effectively saved from property development by Ron Alford, an Axminster farmer who insisted on his grazing rights.

STONEBARROW HILL AND DARTFORD WARBLERS

The outline of the Dartford Warbler is unmistakable

Whereas much of Trinity Hill has been given over to quick-yielding timber under the auspices of the Forestry Commission, Stonebarrow Hill is under the care of the National Trust. Traditional methods are used to ensure that its heather and gorse is never ousted by conifers. In addition, the skillful technique of hedge layering is employed.

This all adds up to an open coastal landscape which suits a great variety of birds, including the Raven and the Peregrine Falcon, species that explore the hill from their nesting sites about the cliffs. The smaller residents include Stonechats, Meadow Pipits, Linnets, and every bird that is likely to turn up in an English garden. But there is a bird that lives on Stonebarrow hill that is worth seeking out above all others: the **Dartford Warbler** Sylvia undata. Part of the hill is managed to suit this rare species.

This perky little bird is our only truly resident species of Warbler, but as a breeding species in England its numbers fluctuate widely according to winter temperatures. It was almost wiped out during the severe winters of the early 1960's. The succession of mild winters that we are now enjoying has ensured a fairly flourishing population of Dartford Warblers in the few sizeable areas of heathland that remain along or close to the Jurassic Coast, such as Aylesbury Common in East Devon and Studland Heath in Dorset.

One or two pairs of these warblers live on Stonebarrow Hill and this is the only place anywhere near Lyme Regis that you a likely to see this species. But the bird is never easy to locate. It spends most of its time flitting about in dense gorse beneath which it nests. It sometimes gives itself away by its rattling song as it perches momentarily on top of a bush, but hardly giving you time to make out its characteristic features. These include the often raised feathers of its greyish-purple crown and and its rather long and frequently cocked up tail.

Access to Stonebarrow Hill by car is up a steep track that rises obliquely off the the eastern end of the main road that runs through Charmouth Village. The turning is sign-posted but easy to miss. There is plenty of parking space on the hill about which a National Trust Information Centre is situated.

THE LAMBERTS CASTLE & THE MARSHWOOD VALE

Several miles to the north of Lyme Regis is a hill topped by an Iron Age hill-fort known as Lamberts Castle. The north-western side of the hill is covered in old Beech. The leaves of this tree form a dense canopy which reduces the amount of light to the woodland floor. Consequently, the growth of shrubs is very much restricted. It is this habitat – woodland with sparse undergrowth – that attracts the **Wood Warbler** *Phylloscopus sibilatrix*, a beautiful summer visitor from Africa.

The bird is easy to identify because of its resplendent plumage. Its white underparts are in striking contrast to its upper sulphur-yellow feathers. Moreover, the bird is easy to locate because of the open nature of its habitat.

Male Wood Warblers arrive in England during late spring. Each one stakes out its claim to part of the woodland before the females arrive a week or two later. The nest is situated close to the ground. The number of these birds visiting England is declining (probably due to problems associated with its African habitat) and your chances of seeing the bird are now slim. And If you would like to record its beautiful song, assuming the birds are about, then you are advised to get up very early, before the sound of traffic which skirts the Beech-wood.

But a visit to the hill is never wasted. The views from its summit are superb and take in several other hillforts that rim the Marshwood Vale. This is a huge area of flat and unspoilt countryside to the east, ideal for a birdwatcher on a bicycle. From the hill-fort its ancient defenders would have seen a dense canopy of Oak, a tree that grows well on the heavy clay soil of the Vale. A little bit of ancient woodland (Prime Coppice) is still left and used to be well known for its Nightingales. From the hill on a clear day you can make out much the Jurassic Coast, including Chesil Beach and the Isle of Portland.

To get to Lamberts Castle take the B3165 out of Lyme Regis and keep going north, crossing the A35 where the Hunters Lodge Inn is situated and following the road to Crewkerne. This road is straight for several miles until it rises sharply as you approach the fort. Slow down and look for an oblique turning to your right which leads to a car park adjacent to the woodland. Nearby is an information panel showing public footpaths and illustrating the birds and other wildlife you might see.

To get up to date with what's happening in the Marshwood Vale pick up a free copy of the Marshwood Vale Magazine. This is available from most news-agents in West Dorset. It is well known for its informative articles on local and natural history. The magazine is published monthly and always includes a top-quality photograph of a local character on its front cover.

The Marshwood Vale is certainly worthy of your attention. Some years ago a by-pass was planned around the village of Chideock (pronounced "Chidick") that fringes part of the Southern perimeter of the Vale. Proposals included an intrusion into the Vale and an alternative scheme that would have cut a swathe through National Trust land to the south of the village. Everyone concerned with the protection of local wildlife was relieved when these projects were shelved.

EXPLORING THE UNDERCLIFF

Most of the six-mile stretch of coast between Lyme Regis and the mouth of the River Axe has been left alone in the wake of massive landslips. What was once a system of open fields bordering the edges of cliffs is now made up largely of an inaccessible jungle full of dangerous crevices. In 1955 it was declared a National Nature Reserve. It is familiarly known as " The Undercliff" or the Landslip.

The cliffs are still on the move and it is dangerous to stray off the public footpath that runs through it. The dominant tree is the Ash. This has a canopy of small leaves through which light easily penetrates, enabling the establishment of a dense undergrowth of shrubs. The Undercliff is full of birds but mostly of the the sort that breed in thick vegetation. Consequently they are very difficult to see, especially during spring and summer when foliage proliferates. Its birds are more likely to be heard than seen. There are open stretches of countryside at either end of the reserve, however, in which many interesting species flit about in full view.

Access to the footpath through the Undercliff from Lyme Regis is from the bottom of Holmbush car-park which is situated on high ground immediately to the west of the town. The footpath borders **Ware Meadows** and leads to **Ware Cliffs** which provide a bird's-eye view of the Cobb and Harbour of Lyme Regis set against a magnificent coastal landscape that takes in much of the Jurassic Coast to the east.

Parts of Ware Cliffs are kept clear by nibbling sheep and rabbits but otherwise the land is made up mostly of tall grasses and shrubs that retain a somewhat tenuous grip on its seaward face which is always crumbling. This is the sort of habitat favoured by many species of warbler as they arrive on our shores from Africa during the spring and early summer. These birds are all about the same size and often look very much alike. If you can see them! They blend in with the foliage within which they remain hidden for most of the time and are more readily identified by their different songs.

These are mostly impossible to imitate by the human voice with the notable exception of the **Chiffchaff** *Phylloscopus collybita.* Its song is like its English name repeated over and over again but is perhaps better rendered by using the German name for the bird – Zilp Zalp. The bird is identical in appearance to the **Willow Warbler** *Phylloscopus trochilus* apart from the colour of its legs (greenish in the "chaff", brown in the "willow"). The bird is not restricted to willow and is common around any sort of woodland. Its song is a melodious descending trill quite unlike that of the Chiffchaff.

The **Blackcap** *Sylvia atricapilla* sports a dark crown and is one of the few warblers that can be told at a glance. And unlike most warblers the sexes of this species are easy to distinguish. The cap of the male is much darker than the female's which is brown rather than black. The song of this bird is a strident warble and is now a familiar sound of the English countryside (It used to be a rather uncommon bird).

The song of the **Garden Warbler** *Silvia borin* is very similar but longer and quieter. Another common warbler is the **Whitethroat** *Sylvia communis*. It lives up to its name and unlike most warblers is often seen perching and singing on the top of a hedge. It never lingers for long in this position and you will have a job to monitor its movements. Its song is loud and scratchy and not unlike that of the **Sedge Warbler** *Acrocephalus schoenbaenus*. As indicated by its name, this species is very much a bird of damp places. It is often heard singing in the dense thickets which shroud a stream passing through Ware Meadows.

Perhaps the most curious warbler you might hear is the **Grasshopper Warbler** *Locustella naevia,* named after its song which, allowing for a good amount of acoustic license, is not unlike the insect in full throttle. This warbler is extremely secretive, always sticking close to the ground amongst dense herbage.This makes it difficult to estimate its numbers but it is generally considered to be a rather rare bird.

The first four species listed are familiar visitors to English parks and gardens but it is in places like Ware Cliffs that they are first seen after leaving their wintering quarters in Africa. Some Chiffchaffs stay here if the winter is mild, but our only truly resident warbler is the Dartford Warbler, the heathland species that we have already tried to find on Stonebarrow Hill.

The laughing chatter of the **Green Woodpecker** *Picus viridis* is often heard over Ware Cliffs. The bird flies from one old tree to another in search of insects that live in the crevices of bark. It sometimes searches the ground for ants and this is the best time to make out its red crown, green back and yellow undersides. In flight the yellow parts of its plumage show up more than the green. A newcomer to bird watching often reports seeing a big yellow bird which turns out to be a Green Woodpecker. In poor light its undulating flight and barrel-shaped body give the bird away.

The coastal footpath over Ware cliffs leads to a tarmacked path which is joined by a lane from the north-west of Lyme Regis and which serves the Crow's Nest, a wooden house situated at the foot of a vertical cliff-face. The path then follows a straight route into the Undercliff proper. It is shrouded by big Sycamore and Ash trees. The tarmac ends about 500 yards from the Crow's Nest where a house is crumbling. This property, known as Underhill Farm, was described in "The French Lieutenant's Woman", a novel by John Fowles that was made into a film.

From then on the path is a narrow and often muddy woodland track kept open only by the tread of walkers and the management skills of Conservation authorities and the East Devon District Council. Most of the birds blend in with their surroundings and the birdwatcher will be disappointed if his aim is simply to get close to the creatures he admires.

The **Treecreeper** *Certhia familiaris,* for example, is common enough about the footpath but you will need a keen eye to make the bird out as it mouses its way up and around a trunk before descending to the ground and making its way up another tree. This tiny bird has a thin curved beak which it uses to pick out creatures that live in the crevices of bark and on leaves.

Another characteristic bird of the woodland is the **Great Spotted Woodpecker** *Dendrocopus major* which, in spite of its large size, about a foot from beak to tail, and its vermilion cap, is difficult to locate.

The **Common Pheasant** *Phasianus colchicus* is abundant and the **Tawny Owl** *Strix aluco* common. Buzzards hover above the canopy of trees and although usually out of sight they are often heard. The voice of a buzzard is not unlike the mewing of a hungry cat. If you hear a piercing shriek then it might will be from a Peregrine Falcon that has missed a Pigeon. The **Marsh Tit** *Parus palustris* and **Bullfinch** *Pyrrhula pyrrhula* are more common than in most woodlands.

It should be remembered that the footpath is merely a cleared thread passing through a coastal belt of 321 hectares of woodland with an exceptionally dense and largely impenetrable undergrowth.The professional ornithologist uses his ears rather than his eyes to determine the species present and to estimate their numbers. He is helped in the knowledge that most woodland birds stick within well defined territories during the breeding season and that bird song usually carries far. Activities unrelated to song also give birds away, notably the drilling of a Woodpecker and the wing clapping of a Wood Pigeon.

Sometimes the woodland is curiously silent apart from the cufuffel of a tumbling squirrel that has lost his foothold on a branch. This might indicate that there is a bird of prey about, especially during early spring when the majority of small birds are normally in full song as they establish their territories. Once the birds are breeding it is to be expected that they will keep quiet as they would not wish to advertise the whereabouts of their nests, especially in a woodland full of squirrels and Magpies that would make a meal out of their eggs and young. That said, the nest of the Great Spotted Woodpecker is a very noisy affair after the chicks have hatched. The youngsters can be heard squawking for food up to a hundred yards away from the hole in the tree in which their parents patiently carved out their nesting site.

Avoid the temptation to stray off the footpath in search of birds.The woodland hides subsiding ground as dangerous as at Black Ven. Perhaps more dangerous. Should you get stuck then you won't be visible to a rescue helicopter flying above the dense canopy of trees even if your clothes are as bright as a parrot. Some idea of the difficulty of finding anyone lost in the Undercliff can be gained from a true story that dates back to shortly after the Second World War. Two young German prisoners escaped from a work party and lived in the Undercliff for two months before they were recaptured. It took a pair of bloodhounds to find them.

You are therefore advised to stick to the footpath, using your ears as much as your eyes as you make your way through the Undercliff. In due course you will encompass the site of a massive landslip that occurred on Christmas Day in 1839. This slip resulted in a chasm about three quarters of a mile long. It is still there, filled up to the brim with huge Ash trees about which dangle the twisted ropes of the Old Man's Beards. This trailing plant and its support grows well on the chalky foundations of the Undercliff. Details of the chasm are difficult to make out because it is so overgrown, but every now and again you will get a glimpse of its inaccessible southern face which provides a safe refuge for breeding birds.

Associated with the chasm is a cliff-top known as Goat Island which is kept clear of encroaching shrubs and trees for the benefit of rare flowers that require a chalky soil. Access to Goat Island is strictly limited to Conservation authorities and is well off the coastal footpath.

The dense woodland eventually gives way to the sunlit slopes of **Haven Cliffs** with fine views of Beer Head in the distance. The slopes are covered in thick scrub, much to the liking of the **Yellowhammer** *Emberiza citrinella*. This beautiful bird is the only species included in this book that has a pleasant sounding Latin name. The reader might well ask why the author has bothered to include such names. It is very much a matter of avoiding confusion. The English "Robin", for example, looks nothing like the "Robin" familiar to every American and which is an entirely different species. But every species has been given a Latin name which is understood throughout the world irrespective of the birdwatcher's native language. And within the British Isles the word "Sparrow" is often applied by the layman to two or three unrelated species as we have noted during our visit to Lyme Regis.

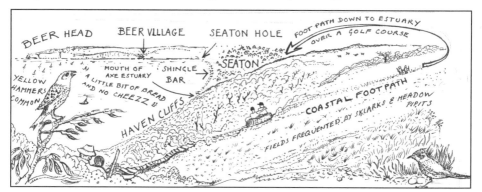

Coming out of the Undercliff

The Yellowhammer is famous for its song which can be roughly imitated by saying "a little bit of bread with no cheese". You will need say each word quickly, one after the other, except for "cheese" which should be drawn out. The bird often leaves this "word" out, leaving its song hanging in mid air. It nests close to or on the ground in dense herbage at the foot of a bush. The bird was very common in England before the days of mechanical hedge flails. These machines rip their habitat apart.

The **Skylark** *Alauda arvensis* is another bird that has seen a drastic reduction in its population. It is a ground-nesting bird of open countryside. Most of the fields bordering the Undercliff have been given over to intensive agriculture and potential nesting sites have been swept aside. Fortunately, some of the fields at the western extremity of the Undercliff are sympathetically managed to suit this species and you have have a good chance of hearing its famous twittering song as the bird rises up in the air. The haunt of the Skylark is favoured by the Meadow Pipit and big flocks of this species are often seen in the fields abutting Haven Cliffs. It is interesting to note that as cliffs erode farmers need to fence off a fair strip of land from the cliff edges. This land might well ensure the survival of small stocks of dwindling species.

The coastal footpath now turns inland and over a golf course before running down to the mouth of the River Axe. We have now reached an estuary full of birds. These are too many to take in after a strenuous walk through the Undercliff and we should allow a full day for their study.

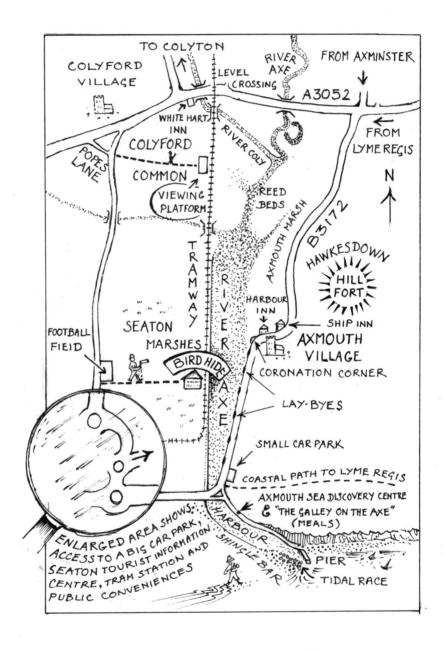

Rough map of the Axe Estuary showing access to bird-watching sites.

EXPLORING THE AXE ESTUARY

Typical birds of the Axe Estuary. This fanciful sketch takes in the village of Axmouth as seen from the western side of the estuary. See text for descriptions of the birds and for historical anecdotes.

To get to the Axe Estuary by car from Lyme Regis head west along A3052. The road is mostly straight over high ground for a good four miles. It then zig-zags down the eastern flank of the valley of the River Axe for about half a mile before meeting a cross-roads at sea level. The road to the left of this junction (the B3172) borders the river as it flows in to the sea and brings you to **Axmouth.**

This attractive village is situated about a bend in the road and nearby is a grassy triangular patch called **Coronation Corner**, named in celebration of the Jubilee year (1973) of our Queen's coronation. This little picnic area commands splendid views of Axe and its birds as the river broadens out into as an estuary. It is especially suited to the needs of a bird-watcher who can't walk far: Coronation Corner is adjacent to the Harbour Inn and its parking facilities, and close by is another pub, the Ship Inn, run by the Chapman family for over thirty years and renowned for its fine food. It houses an aviary for sick and injured owls which can be viewed by visitors to the inn.

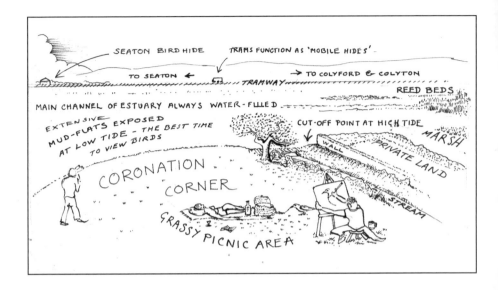

A small stream runs though the village and flows into the estuary at Coronation Corner. It can easily be crossed at most states of the tide, enabling you to explore the soggy marsh to the north. (You are advised to wear Wellington Boots).

The marsh is often the haunt of the **Snipe** *Gallinago gallinago,* a secretive bird whose variegated brown and grey plumage blends in with its habitat. Its bill is long and straight, well suited for digging up the worms that form a large part of its diet. Several individuals might be about without your knowledge. If you get too close to the flock then one or more of birds will suddenly take to the air with a characteristic zig-zag flight, keeping low before landing somewhere else in the marsh. The Snipe is a good example of a bird that uses both camouflage and behaviour to confuse its enemies.

Raising a Snipe.

The mouth of the River Axe was formerly much wider than it is today and for many centuries ships unloaded their cargo in the vicinity of the village. It is believed that Axmouth was a Roman port situated at the foot of the Fosse Way, the Roman frontier established by the future Emperor Vespasion in AD 47 and which stretched diagonally across England somewhere in the vicinity of Seaton to the north-eastern coast of Yorkshire. Vespasion was a friend of Pliny the Elder who wrote several volumes on Natural History. Perhaps these great men discussed the bird life that Vespasion saw as he relished the thought of extending his Empire to the west of the River Axe.

Any evidence of Axmouth as a Roman port is buried beneath its streets and its muddy perimeter, long awaiting the attention of an archaeologist rather than an ornithologist. But we can be sure that in the days of the Roman Empire the sort of birds and other wildlife we see about the Axe today were very different. We can imagine Vespasion's Axe as a sort of River Amazon, carving its way through a landscape of dense forest, the haunt of bears, beavers and vultures, birds that might well have picked the bones of the defeated tribes that built the numerous Hill Forts that encompass the Dorset/Devon coastal border.

Axmouth was described as an important fishing town with sixteen inns by a visitor during the seventeenth century. Some time afterwards the mouth of the estuary gradually silted up in the wake of a developing bank of pebbles across its entrance, a process that can be attributed to cliff erosion to the west of the Axe. This is still going on and the mouth of the estuary is now a narrow tidal race hugging the eastern cliffs about one mile downstream from the village.

The road between Axmouth and the harbour is dead straight with several lay-byes along its course. These are good stopping points for viewing birds, especially when the tide is out and when extensive mudflats are exposed. The mud is black and unappealing to a holiday-maker in search of sand but good news for birds. It contains countless millions of creatures that provide a vast supply of food which the birds gobble up. Moreover, these birds can feed unmolested by predatory mammals.

The mud is also good news for the birdwatcher because it provides a uniform background against which many birds stand out. This is in contrast to the situation on a rocky seashore or a pebble beach where, as we have seen, most wading birds blend in with their habitat and are difficult to detect unless they take to the air.

Several species that utilise the mudflats can be recognised from their silhouettes alone. Of these species the **Curlew** *Numenius arquata* is an outstanding example. Its beak is curved and about a foot long, making the species easy to identify at a great distance as it probes the mud on its long, stilt-like legs. It gets its name from its characteristic call.

The Curlew is often seen burying its beak in mud up to its eyes but it is not easy to make out exactly what it is feeding on, even with use of binoculars. Like most birds of its kind it appears to engulf its food before lifting its beak out of the sediment. The Curlew is a big bird and no doubt goes for the larger creatures that burrow fairly deep within the mud.

Of these creatures only two species are obvious, at least to an angler digging for bait: a two-shelled mollusc called the Peppery Furrow Shell and a worm no longer than your little finger and familiarly known as the Harbour Ragworm. (The huge King Ragworm, growing up to three feet long and as thick as a finger, is absent from the Axe estuary).

The Curlew could be mistaken for a **Whimbrel**, *Numenius phaeopus*, a rather uncommon bird of passage. This bird is slightly smaller than the Curlew and its beak is less curved. Otherwise, both species sport a similar mantle of brown mottled feathers that make them difficult to see when they forage in the marshes after the tide has come in. If you see what appears to be a Curlew with a straight beak then the chances are that it is a species of **Godwit**. The legs of Godwits, particularly those of the **Black-tailed Godwit** *Limosa limosa* (nicknamed "Blackwit"), are very long in relation to their bodies and poke out behind their tails when the birds are flying.

Of the smaller waders that live on the estuary, the **Redshank** *Tringa totanus* is especially distinctive. Its beak is straight and its legs are scarlet, a feature that gave rise to its common name. It has a distinctive call which can be rudely imitated by repeating the words "clue you". These notes together with the somewhat similar ones made by the Curlew and the "kleep kleep" of the Oystercatcher are largely responsible for the pleasing repertoire associated with an estuary throughout the year.

The Redshank is a common resident in contrast to its visiting relative the **Greenshank** *Tringa nebularia* which as indicated by its name has green legs.

Going down in scale, there are host of small species of waders that are not easy to tell apart without the aid of binoculars.The commonest of these species is the **Dunlin** *Calidrus alpina*. It is about the same size as a starling. The smallest wader of all is the **Little Stint** *Calidris minuta*, a really tiny bird no bigger than a Robin but nothing like it in profile. It is best appreciated through a telescope.

The smallest British wader, the LIttle Stint, with
one of the largest, the Black-tailed Godwit.

The B3172 bears sharply to the right over a bridge as the road approaches the sea. The coastal footpath from Lyme Regis through the Undercliff ends just before this bend where there is parking space for several cars. This is where a bus-driver from Lyme Regis drops off passengers who intend to walk back to the town through the Undercliff. It is also a good place for the bird-loving motorist to stop if only to feed the ducks, pigeons and gulls that have got used to picnickers discarding bits of their sandwiches on the nearby shore.

Cormorants are often seen fishing in the vicinity of the bridge and there is always a good chance of seeing a Kingfisher. But the constant traffic scares the rarer birds away and to get the best of your birdwatching you will need to cross the bridge and explore the other side of the estuary.

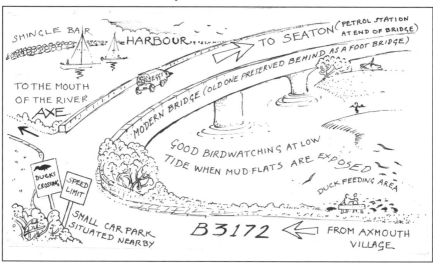

After traversing the bridge the road runs parallel to Seaton sea-front for about a quarter of a mile. A large car park is situated off the first roundabout you come across. Nearby is the **Seaton Tourist Information Centre** and the **Seaton Tramway Station**. You will not be disappointed if you take a ride. The tramway links up Seaton with Colyton, an ancient (Anglosaxon) town about two and a half miles to the north, and follows a straight route along the estuary, bordering the edge of reclaimed land to the west. The trams provide a useful service for local residents and are popular with tourists. There could hardly be a better way to take in the grandeur of the Axe valley.

For the archaeologist there is the thrill of knowing that he is travelling in the middle of what was once a wide estuary where Roman ships used to sail. For the ornithologist there is the opportunity of getting close to estuarine birds, most of which are naturally shy. The trams double up as mobile hides and special trips are available to birdwatchers. These excursions are normally held during the colder months of the year when the estuary is swarming with overwintering birds. (Contact the Seaton Tourist Information Centre for details). If you take lunch in Colyton then you then you should find a local inn, the The Kingfisher, very much to your taste!

The trams pass within a couple of metres from **The Seaton Marshes Bird Hide.**
This outstanding facility is free of charge and open throughout the year. For much of
the time it is occupied by local ornithologists who will gladly give advice whether you
are a novice or an expert birdwatcher unfamiliar with the area. From time to time an
exhibition of binoculars and telescopes is put on within the hide by a leading retailer,
giving you an opportunity to try out the latest equipment.

The approach to the hide is wheelchair friendly and consists of a flat path hidden
by embankments and out sight of birds that would otherwise be upset by your pres-
ence. It is important to stick to this tract in order to avoid disturbing a nearby pond
that has been cordoned off for breeding birds.

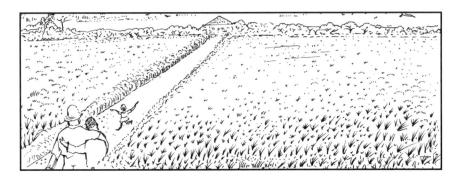

On the way to the Seaton Marshes Bird Hide

A bird table is situated within a few feet of the hide, close to the bushes that flank
the tramway. These bushes are frequented by many common resident species such
as the Greenfinch *Chloris chloris* and Chaffinch *Fringilla coelebs*. During the spring
this vegetation provides refuge for exhausted warblers and the like that have migrat-
ed from Africa. The various species and the dates of their arrivals are eagerly chalked
up within the hide by local bird-watchers.

Inside the bird hide

The bushes extend along the sides of the tramway and are the haunt of the **Sparrowhawk** *Accipiter nisus*. Do not be surprised if you see this bird swoop on to the bird table and carry off a finch in its beak. Sensitive people are often dismayed at the sight of a bird of prey capturing its food and have sometimes to be convinced that the life of the victim is over in a matter of seconds. Nature really is "red in tooth and claw" as Alfred Lord Tennyson put it. And birds of prey would not be around unless all is well in Nature. They are at the top of a food chain and dependant on a flourishing population of their food supply.

The Sparrow Hawk is a common British bird but unlike the Kestrel and Buzzard it keeps low and is rarely seen. When hunting it goes for a quick kill, immediately returning to the bushes in which it hides and where it has a favourite perch on which it devours its prey. In human terms the Sparrow Hawk is cunning. It is much more aware of us than we are of it. The bird might well swoop over your shoulder without you noticing. It has even been said to fly into a local pub in search for a pet canary, an event that can not be verified.

The most colourful bird that hunts in the vicinity of the hide is the **Kingfisher** *Alcedo atthis*. It is a contender for Britain's loveliest species. Throughout the year the estuary and its tributaries are full of the small fish that make up much of the Kingfisher's diet. From about the end of April the fish population of the Axe is swelled by mullet returning to the estuary after spending the winter out to sea. These fish grow big and are often seen foraging about the mud with hardly any water covering their backs.

Many were scooped up by an **Osprey** *Pandion haliaetus* which lingered on the estuary for a week or two during the summer of 1995. This rare and magnificent passage migrant was the talk of the hide and its picture featured in the national press. Since then an Osprey has been seen on the estuary almost every year.

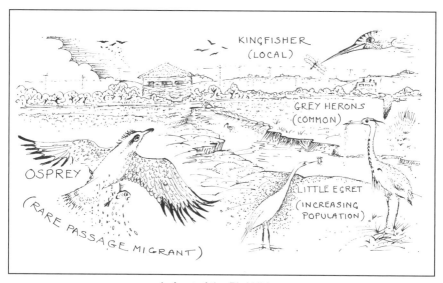

In front of the Bird Hide

The **Grey Heron** *Ardea cinerea* is often seen stalking for fish at the waters edge but it is not as common as it used to be. We learn from the fourth edition of Pulman's famous Book of the Axe, published in 1875, that the ancient deer park associated with Shute House (situated about four miles to the north of Seaton) "formerly contained a heronry, omitted in Mr. Yarrell's list of heronries. But about twenty-five years ago the herons left their old home and adopted a new one among the woods of Stedcombe near the mouth of the river". The Stedcombe heronry was felled at the end of the !970's. A smaller one was established in a clump of conifers in a field immediately to the north of Axmouth village. The trees were damaged in the year 2000 and the herons now roost in the woodland that clothes the foot of Hawksdown Hill.

The history of the **Little Egret** *Egreta garzetta* is in the reverse order.This beautiful snow-white member of the heron was a very rare visitor to the British Isles up until the mid 1970's when its numbers noticeably increased. By the late 1990's it was established as a British breeding species. It is now seen always to be seen on the Axe estuary, with numbers approaching forty during a recent winter. The Little Egret has yet to breed here, however.

An attractive feature of herons is the array of fine feathers that droop from the back of the head in many species.The feathers that make up this crest are especially beautiful in the **Snowy Egret** *Egretta thula*, an American species that has never been recorded in Britain or Europe. In the days of Audubon these feathers were literally worth twice their weight in gold and the bird was hunted to the point of extinction. It was saved by the efforts of the National Audubon Society which adopted the bird as its logo.

There is always something special to look out for on the Axe estuary whatever the time of the year. Monthly records of the species seen are posted up along the path leading to the hide and also in the hide itself. They often include species of great rarity. An estimate of the numbers of each species is also included in the reports. These numbers are reckoned in the order of thousands for some species during the autumn and winter when birds from elsewhere hone in to the mud flats for food and security. Of these species two are very evident: the **Lapwing** *Vanellus vanellus* (a type of plover) and the **Widgeon** *Anas penelope* (a duck).

The Lapwing lives up to its name. Its rectangular black and white wings lap about when the bird takes to air in the manner a giant butterfly. At the same time the bird often utters a call from which it gets another common name -"pee-wit". When roosting on the mud flats at low tide a flock of lapwings lines up like an army with their beaks usually pointing towards the mouth of the estuary. And like an army, each bird sports a clear-cut uniform: glossy dark shoulders which contrast with a white breast and a fine headgear which consists of a helmet of wispy feathers that project from the back of its head. These beautiful birds never seem to do much while they line up in formation on the mudflats.They are content in the day-time security of their location, patiently awaiting nightfall when they retreat to the marshes and farmland to feed.

An "army " of Lapwings

Unlike the Lapwing, the Wigeon is not very easy to make out as a distinctive species from a distance, especially when seen against the sun. Its shadowy profile then resembles that of most ducks and you will almost certainly need binoculars to make out the attractive plumage of the drake. He sports a yellow crown over his chestnut head and the white parts of his grey wings are distinctive in flight. Its rear end is black. The female, as in most ducks, is brownish all over apart from a white patch on her wings. This patch, known as a **speculum,** is a useful feature in telling ducks apart. Its colour varies from one species to another as your Field Guide will testify.

The Wigeon is a truly wild duck and you will never get close to it in contrast to some **Mallard** *Anas platyrhynchos* a species from which most domesticated ducks are derived. But is only the progeny of tamed mallards that waddle up to our feet in pursuit of food. Those hatched in the wild are flighty as Wigeon. The speculum of the Mallard is bright blue.

All sorts of wild duck utilise the Axe Estuary and many of them are very colourful. Perhaps the most attractive species is the **Teal** *Anas crecca* which is unmistakable because of the green dolphin-shaped patches that adorn each side of its brick-red head.These patches are accentuated by a thin yellow stripe that runs around them.The speculum of this species is green.

Several species of duck can recognised by the shapes of their beaks alone. A classic example is the **Shoveler** *Anas clypeata* a bird that lives up to its common name. It uses its broad and flattened beak to scoop up food at or just below the surface of the water.

All these birds are examples of "dabbling ducks". They often "up-end" as they stretch their necks beneath the water but otherwise they stick to the surface in contrast to "Diving ducks" which completely disappear in pursuit of food. Such birds include the **Pochard** *Aythya ferina* and the **Tufted Duck** *Aythya fuligula* but these species are never common on the Axe Estuary.

Grebes are often mistaken for ducks but are not related to them. The **Little Grebe** *Podiceps ruficollis* (also known as the **Dabchick**) is an expert diver and is common on the estuary during the winter. Pulman desribed this bird as a "curious lump of adipose tissue".

The **Shelduck** *Tadorna tadorna* is a big and common bird of the estuary. It is very easy to recognise because of its clear-cut plumage of orange, white and black. It does well locally and you are almost certain to see a brood of Shelduck chicks following its parents during the summer if you visit the hide or take a tram trip. It makes its nest in a tunnel in a mud bank, typically a hole vacated by a rabbit.

A family of Shelduck

Ducks are closely related to swans and geese. The **Mute Swan** *Cygnus olor* is familiar to everyone living in the British Isles. A number of pairs nest about the Axe Estuary and flocks of over forty visiting birds have been recorded locally.

The **Canada Goose** *Branta canadensis* was brought to England as an ornamental bird in the seventeenth century. It has since spread all over the country and is common on the Axe Estuary during the winter. The bird looks as if a black stocking had been drawn over its head and down its long neck. The "stocking" is split around the cheeks where white patches show through. This species could be mistaken for other geese with a similar attire but the backs of these birds are either dark grey or black in contrast to the brown upper parts of the Canada Goose.

Canada geese

Many birds of the estuary spend much of the day in reed beds. Typical examples include members of the rail family of which the **Moorhen** *Gallinula chloropus* and **Coot** *Fulica atra* are well known. They are often seen on ornamental lakes and ponds where the birds have got used to human company. In contrast, the rarer **Water Rail** *Rallus aquaticus* always remains secretive and is difficult to locate because its plumage blends in so well with reeds. Rails such as the Coot might be confused with ducks as they swim very much like them. The feet of rails are not webbed like those of Ducks, however.

Several birds typical of reed beds have been named after this habitat, notably the **Reed Bunting** *Emberiza shoeniclus* and the **Reed Warbler** *Acrocephalus scirpaceus*. Both species live on the Axe Estuary.

The reed beds are not extensive enough to support the **Bittern** *Botaurus stellaris*. This extraordinary bird depends entirely on reed for its survival. Its plumage and configuration blends in perfectly with its environment and the bird mimics a clump of reed by stretching its neck and pointing its beak to the sky. At the same time it swivels its eyes downward, looking out for danger. The bird is famous for its booming call that carries a mile or more. This ability is associated with its very long windpipe which curls up beneath its breast bone and acts as a resonating chamber.

A variety of unrelated birds utilise the Axe Estuary. Included here are rails (Moorhen, Coot and Water Rail) ducks (Mallard, Shoveler, Widgeon, Pintail and Teal) warblers, a bunting, and the very rare Bittern.

We can do no further justice to the great variety of the birds seen on and around the Axe Estuary. And to get up to date with the latest sightings you will need to visit the bird hide where a daily record is kept. But all the species that have been recorded in recent years together with estimates of their populations have been the subject of a useful report* by David Walters, the voluntary recorder at the reserve.

He concludes " that in this small area of less than 3 square kilometres, about 225 species (14 of which are Devon County Rarities) have been recorded, which represent over 39% of the 568 species on the British List. Perhaps there are other small areas that can boast more, but we are indeed well endowed here."

*Axe Estuary Bird Report. Compiled by David Walters, January 2005. Available at modest cost from its author (Address: 7 Springfield, Colyford, Colyton, Devon EX24 6RE).

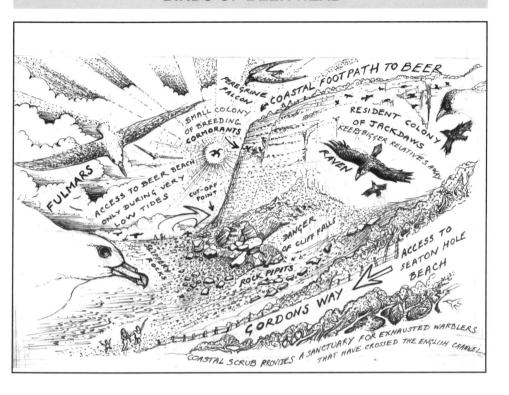

The headland between Seaton and the village of Beer is faced with sheer cliffs, mostly of chalk, which provide a sanctuary for a variety of coastal birds. Most of the species can be viewed from the sea-shore at Seaton Hole, the name given to the small bay at the western end of Seaton Beach where the headland begins.

At present, access to Seaton Hole is very convenient for the motorist who has only to follow the coastal road out of Seaton, keeping left all the time and looking out for the appropriate sign post. There is plenty of parking space along the road above Seaton Hole together with a cafe and toilet facilities and close to which is a footpath down to the beach. This path is called Gordons Way, named after the local hero who kept it open in spite of cliff-falls. There is no guarantee that the path will survive, however.

The Raven and Peregrine Falcon are often seen flying about Seaton Hole. **Jackdaws** Corvus monedula are very common and make their nests in crevices situated high up in the cliffs. The headland also supports a small breeding colony of Cormorants during the spring and early summer. The colony can be viewed from above through breaks in hedges along the coastal footpath.

The **Fulmar** *Fulmaris glacialis* glides about the edges of the cliffs during the warmer months of the year.

At first sight this bird looks very much like a Herring Gull but with stiffer wings. The Fulmar is not a gull, however. It belongs to the albatross family. A characteristic feature of each member of this family is a system of tubular horny sheathes investing the beak. These are difficult to make out in the Fulmar without the use of binoculars. The bird spends most of its life out at sea and it is only in recent years that it has become common around the English Coast. The reason for its spread, like that of the Little Egret, is something of a mystery.

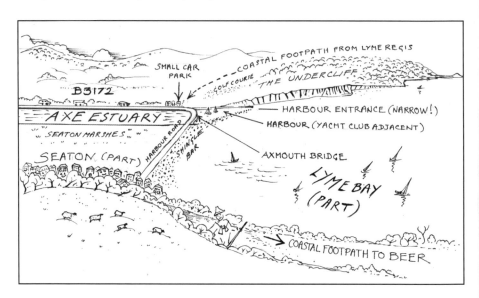

Looking east from the coastal footpath above Seaton Hole

The coastal walk along the cliffs above Seaton Hole to Beer is good for birds at any time of the year and the scenery is superb with outstanding views of a lot of the coastal countryside that we have explored.

The seascape takes in much of Lyme Bay. Many oceanic birds frequent the Bay and with this in mind we return to Lyme Regis and seek out the services of local boatmen.

BACK TO LYME REGIS: TAKING A BOAT TRIP

Lyme Regis is situated near the middle of the coastal arc between Start Point in Devon and Portland Bill in Dorset. This arc is the boundary of Lyme Bay, an area frequented by many birds that spend most of their lives at sea. These oceanic species include birds that form big nesting colonies on islands or on inaccessible cliffs where they are undisturbed by human activities.

The largest of these species is the **Gannet** *Sulla bassana*. It gets the second word of its Latin name from Bass Rock, a small rocky island off the east coast of Scotland. This island supports a colony of Gannets that has flourished for centuries. The birds plunge dive for fish.

The breeding quarters of the Gannet are shared by colonies of smaller but equally fascinating birds,notably the **Puffin** *Fratercula artica* and the **Guillemot** *Uria aalge*. These species dive from a sitting position on the surface of the sea. They are sometimes brought close inshore after stormy weather and occasionally drift into the harbour of Lyme Regis.

To stand a fair chance of seeing any of these species you will need to take a boat trip. This facility is readily available from about Easter to the end of September. Skippers advertise their services on the railing opposite the life-boat station on the Cobb. The trips are mostly geared up to one hour inshore excursions with the option of mackerel fishing on the way. Your skipper will be happy to point out any oceanic birds that are about.

Boat trips further out to sea are also on offer. These are geared to deep sea anglers and it is during these trips that gannets are most likely to be seen. You might also glimpse a **Shearwater** or a **Petrel** while **Skuas** fly higher and pursue any bird that has been successful in capturing fish.

If you are a member of a bird-watching group then consider booking a boat for your party. Arrangements can be made by chatting to the boatmen or by contacting the Tourist Information Centre of Lyme Regis. All the skippers are ornithologists in their own right and will tell you about the latest sightings of birds out to sea.

Thousands of birds cross the English Channel during the spring and autumn migratory periods and the smaller species often flop exhausted on boats – a good talking point when you go to sea with a Lyme skipper at the helm!

Oceanic birds are often the victims of oil slicks. Fortunately, there is growing public awareness of the need to ensure that our seas are kept free of pollution. And it is gratifying to note that all the birds that we have sought now live within a World Heritage Site.

If the ghost of John Gould is about then he would surely approve.

FOSSIL HUNTING
AROUND
LYME REGIS

A PRACTICAL INSIGHT INTO
THE JURASSIC PERIOD

Written & Illustrated by
DR COLIN DAWES BSc PhD

By the same author and in the same series
"Written in a clear and informative style, it begins with an explanation
of just why there are so many fossils in the area and an overview
of Jurassic sea life. It is filled with clear diagrams and drawings,
and makes even the most complicated geology accessible".
The Marshwood Vale Magazine
(for West Dorset, South Somerset and East Devon).

ISBN 0-9520112-1-2